Bee Bee's Circus

Number Fun Activity Book

by Jackie Reynolds

Dimpled Hippo

To LIAM !! :)

Keep on Counting!

♡ Jackie Reynolds

"Bee Bee"

2015

For my mom, with love.

Copyright © 2015 by Jackie Reynolds

ISBN-13: 978-0-9908357-1-4
LCCN: 2014918703

Printed in the United States of America

Attention schools, libraries, preschools & homeschool organizations: Quantity discounts are available on bulk purchases of this book for educational, fundraising or gift giving purposes.
For information please contact:
Dimpled Hippo, Jackie Reynolds, PO Box 207, Red Hook, NY 12571.
(845) 758-1938, or send an email to: orders@beebeetheclown.com
Visit: www.DimpledHippo.com and www.BeeBeeThe Clown.com

We can count backwards from 10 down to 1.

Bee Bee's Circus Number Fun

Color the circles as you count the things at the circus.

You can count backwards too!

Use your imagination to make your own circus!

Try using a stamp pad with a stamper or fingerprints to color the circles.

Write your name.

Neigh
Neigh

Find the bouncy bows.

Ten pretty ponies with ten bouncy bows
run in ten circles and dance doe-see-does.

Maaaa
Maaaa

I can count upside-down.

Which is your favorite goat?

Nine shaggy goats with nine hairy chins
nibble nine sweet-n-salty popcorn tins.

Let's count backwards:
8, 7, 6, 5, 4, 3, 2, 1

Write the number eight. 8

Spin Spin

I don't wear shoes.

What color are your shoes today?

Eight fancy dancers with eight pairs of shoes
spin eight pirouettes in frilly tutus.

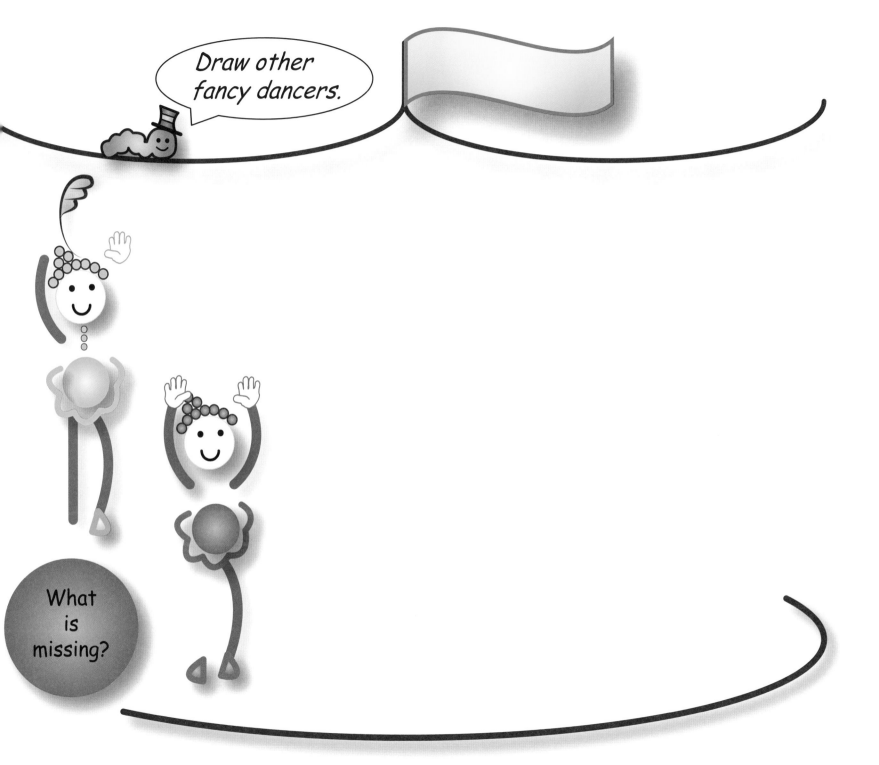

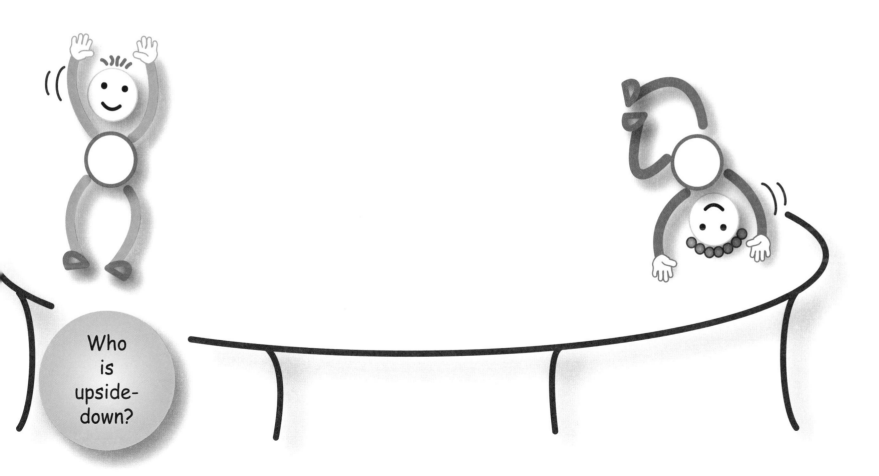

Let's count backwards:
6, 5, 4, 3, 2, 1

Write the number six.

Six country cows in six different sizes
win six rosette ribbons for Counting Fair prizes.

Connect the animals with their sounds.

Start with a circle and draw your own animal.

5

Ruff Ruff

Dance a happy dance.

I dance and wiggle!

Five prancing poodles with five poufy tails
dance five jitterbugs with glittered toenails.

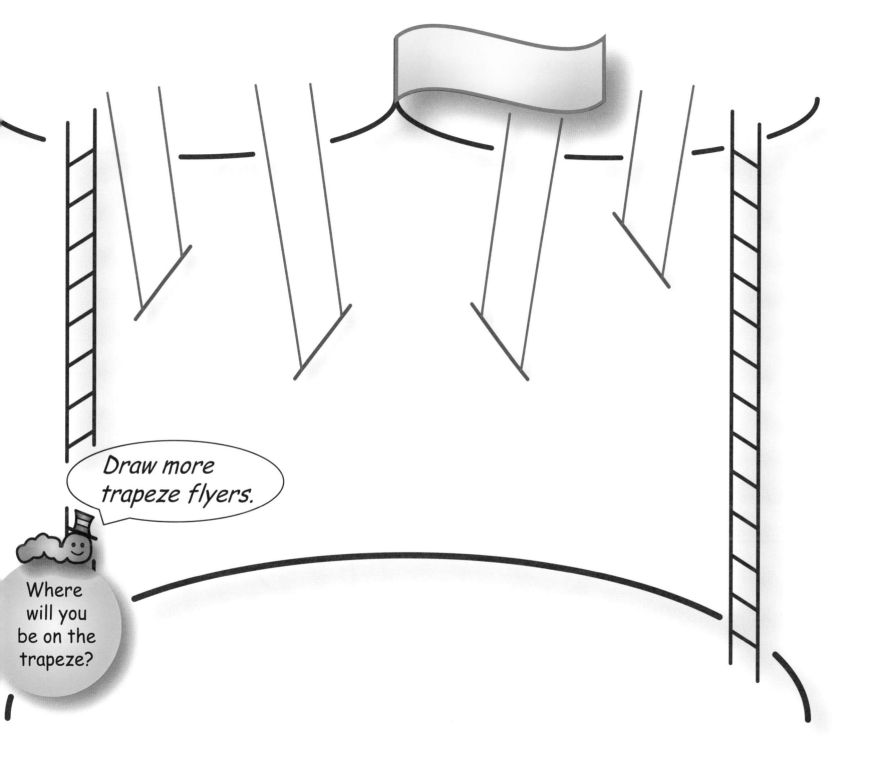

Let's count backwards:
3, 2, 1

Write the number three. 3

3

Vroom!
Vroom!
Vroom!

3

Which motorcycle is upside-down?

2

1

I count going down.

Three motorcycles go *Vroom! Vroom! Vroom!*
around and around in a rumpus room.

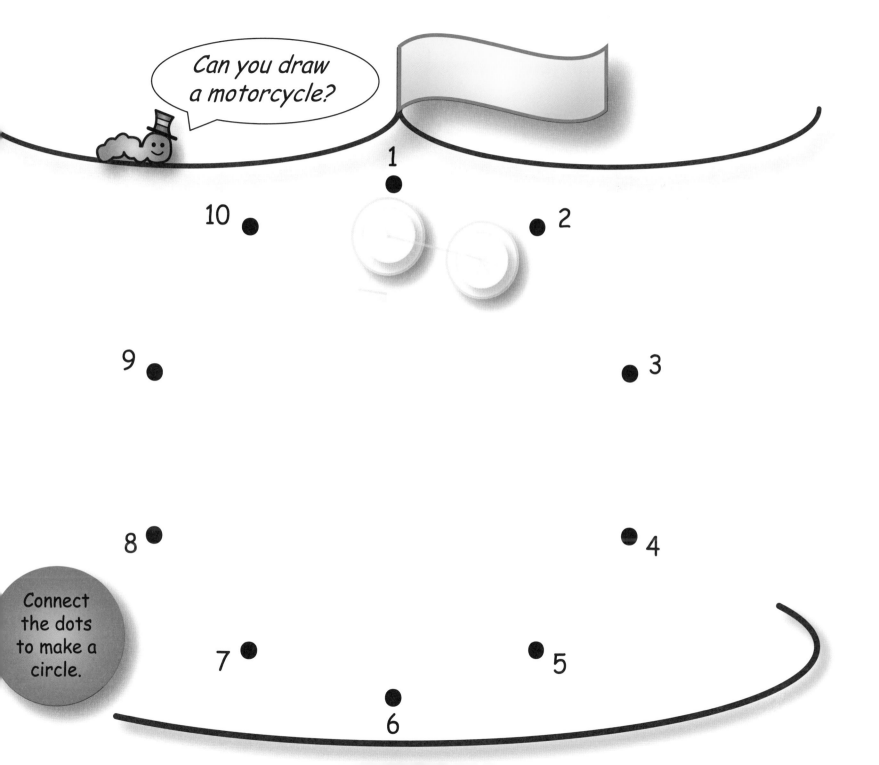

Let's count backwards:
2, 1

Wheeee!

Write the number two.

2

Try saying:
"Pick a Pickle"
2 times fast.

Oink Oink

2

1

Two silly clowns with two spinning plates
race two plump pigs on pink roller skates.

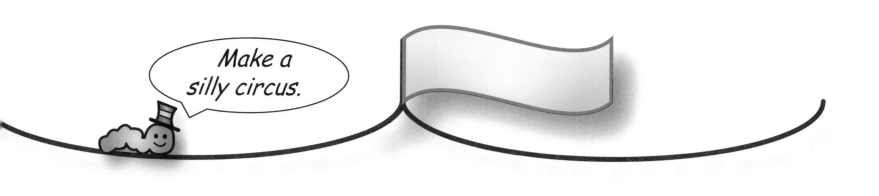

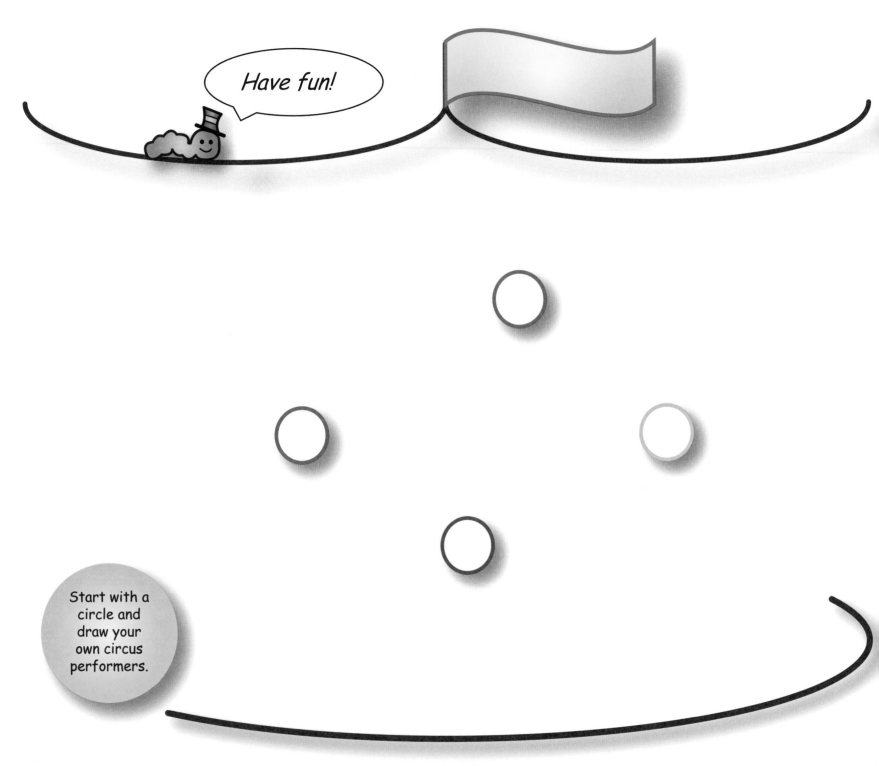

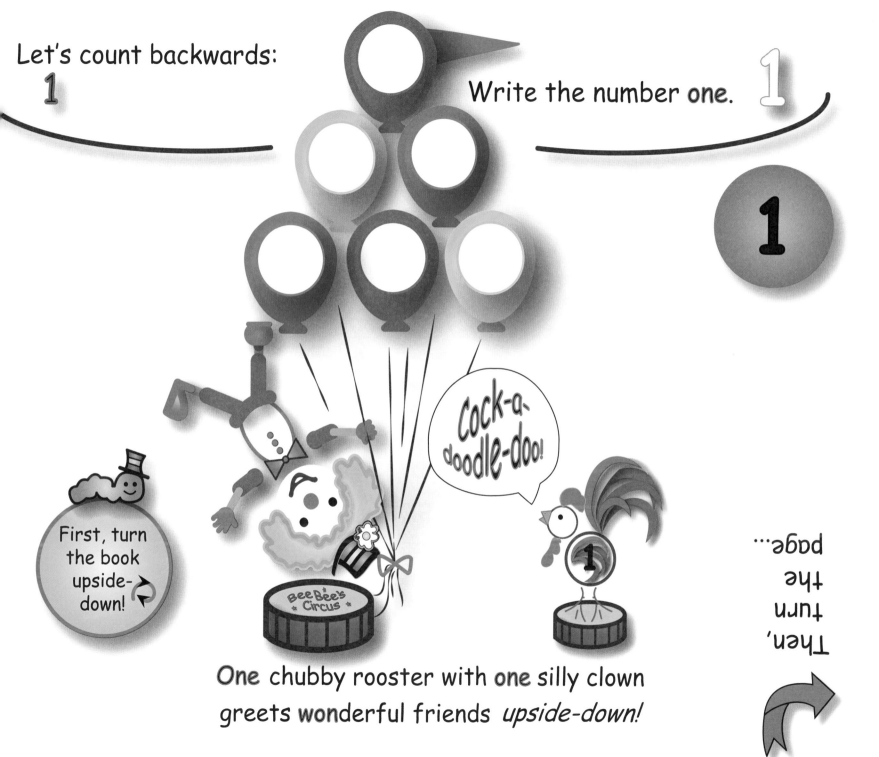

Let's count backwards:
1

Write the number **one.**

1

1

Cock-a-doodle-doo!

First, turn the book upside-down!

One chubby rooster with one silly clown greets wonderful friends *upside-down!*

Then, turn the page...

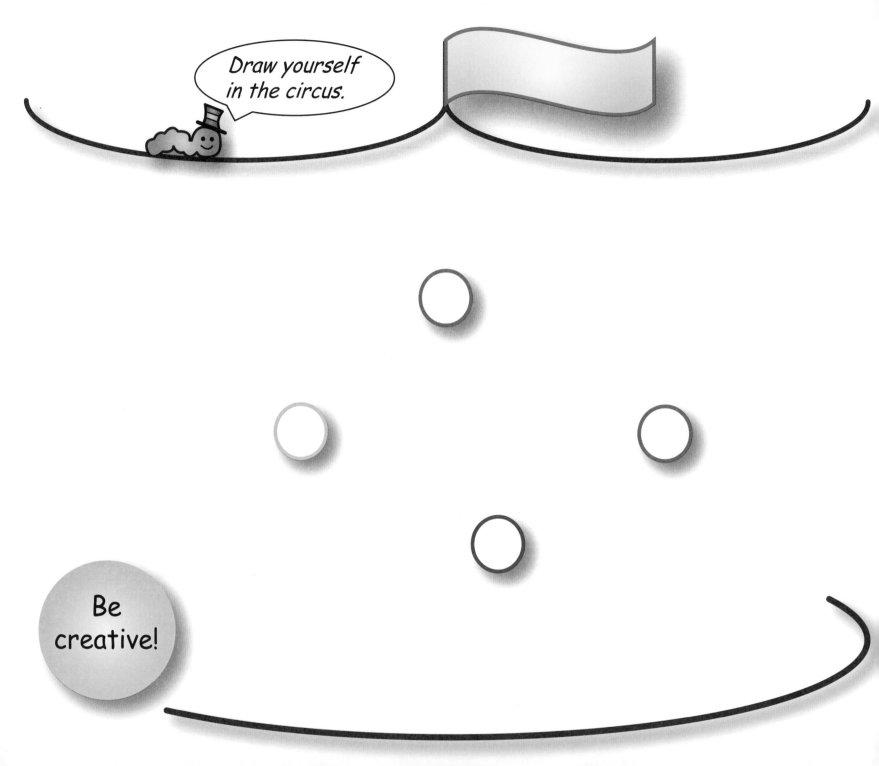

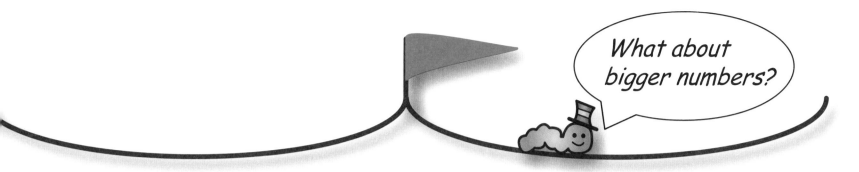

A BIG drum roll, please...

Let's start with **20** and count backwards...

Many happy children with twenty cotton candies

20 19 18 17 16 15 14 13 12 11

and nineteen popcorns

19 18 17 16 15 14 13 12 11

and eighteen balloons

cheer as...

18 17 16 15 14 13 12 11

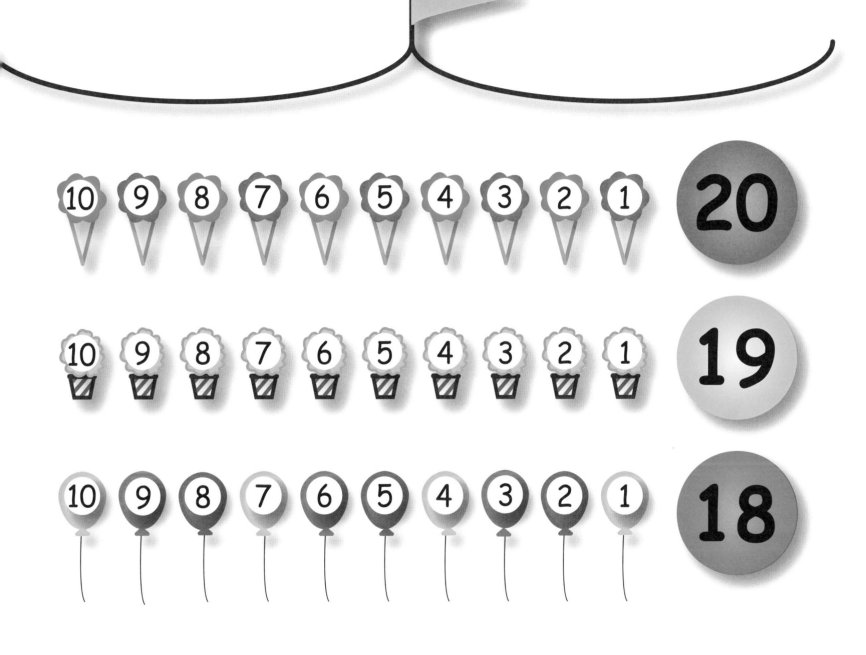

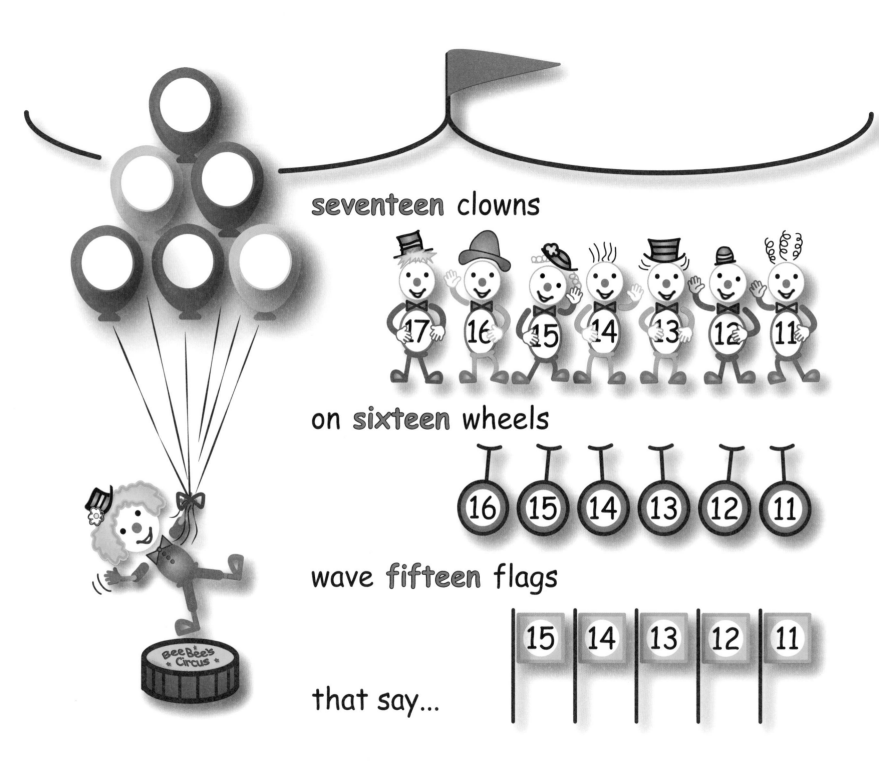

seventeen clowns

on sixteen wheels

wave fifteen flags

that say...

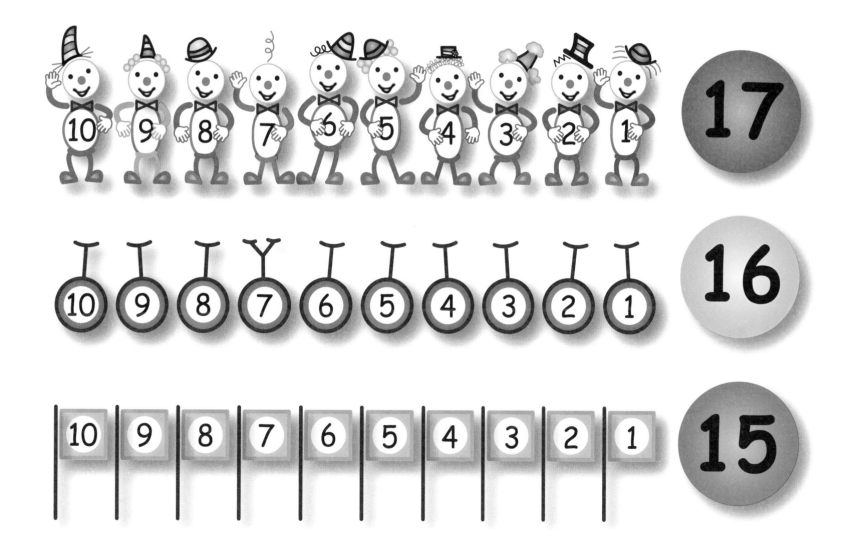

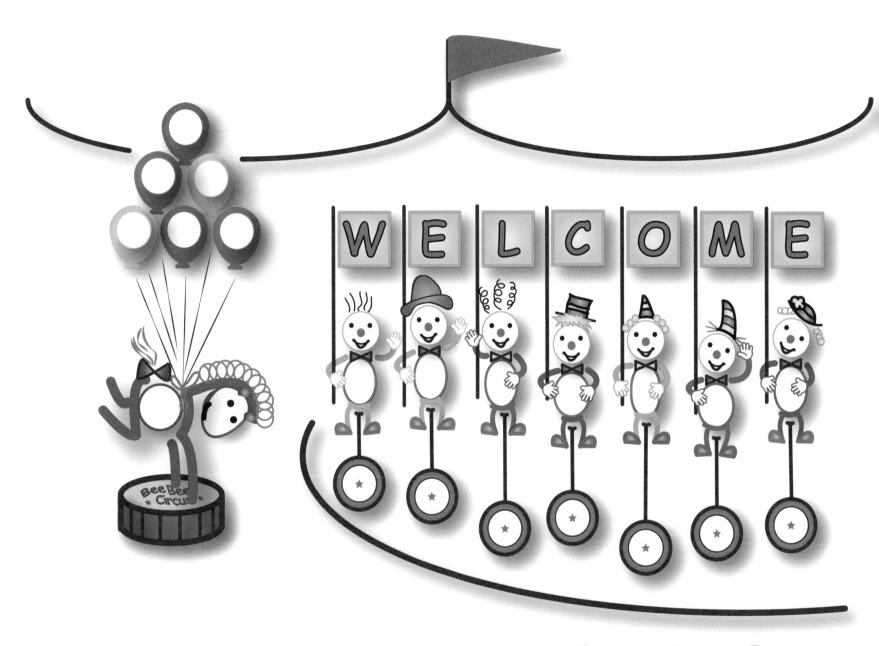

WELCOME

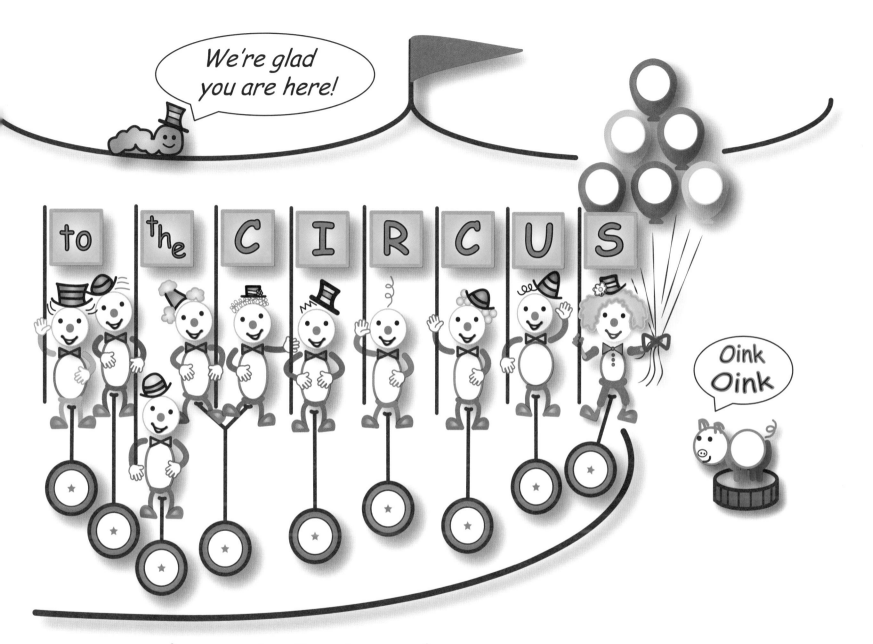

to the CIRCUS!

Lots of kids, with **fourteen** lights, laugh

as **thirteen** clowns

toot **twelve** tin horns

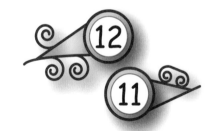

and play...

Draw your family at the circus.

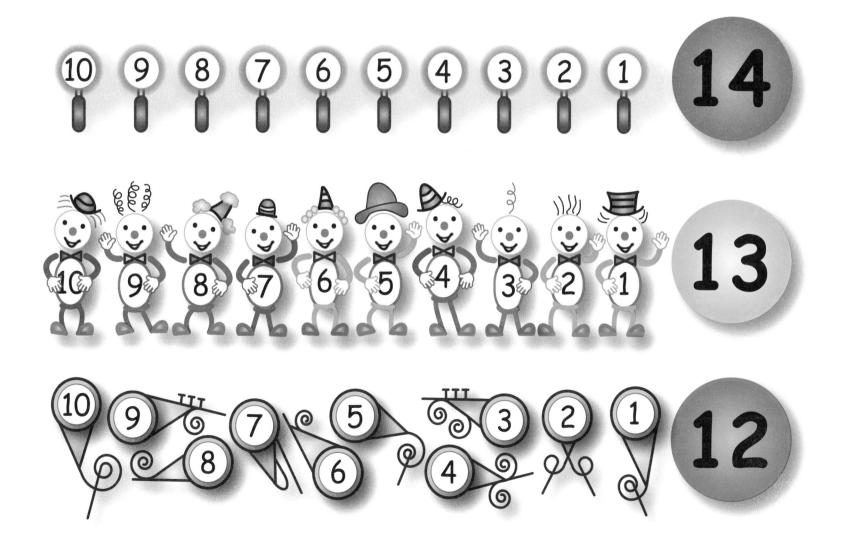

Finish the Show! Count from 10 down to 1.

Floop
Boink Whizzz
Zip Bonk Zing
Bloop Twang Ping
Pop Peep

11

What is your favorite silly circus sound?

eleven silly circus sounds!

Let's count backwards:
10, 9, 8, 7, 6, 5, 4, 3, 2, 1

Let's count together!

English	Spanish	Japanese
10 ten	10 diez	10 juu
9 nine	9 nueve	9 ku
8 eight	8 ocho	8 hachi
7 seven	7 siete	7 shichi
6 six	6 seis	6 roku
5 five	5 cinco	5 go
4 four	4 quatro	4 shi
3 three	3 tres	3 san
2 two	2 dos	2 ni
1 one	1 uno	1 ichi

10

1

We can count in many languages.

20	twenty	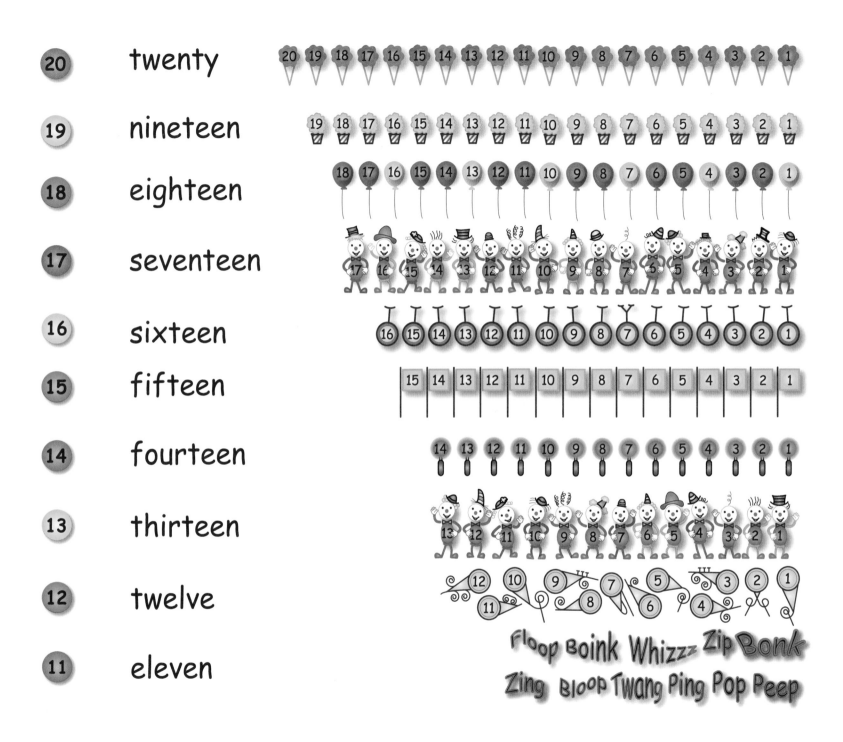
19	nineteen	
18	eighteen	
17	seventeen	
16	sixteen	
15	fifteen	
14	fourteen	
13	thirteen	
12	twelve	
11	eleven	

Floop Boink Whizzz Zip Bonk
Zing Bloop Twang Ping Pop Peep

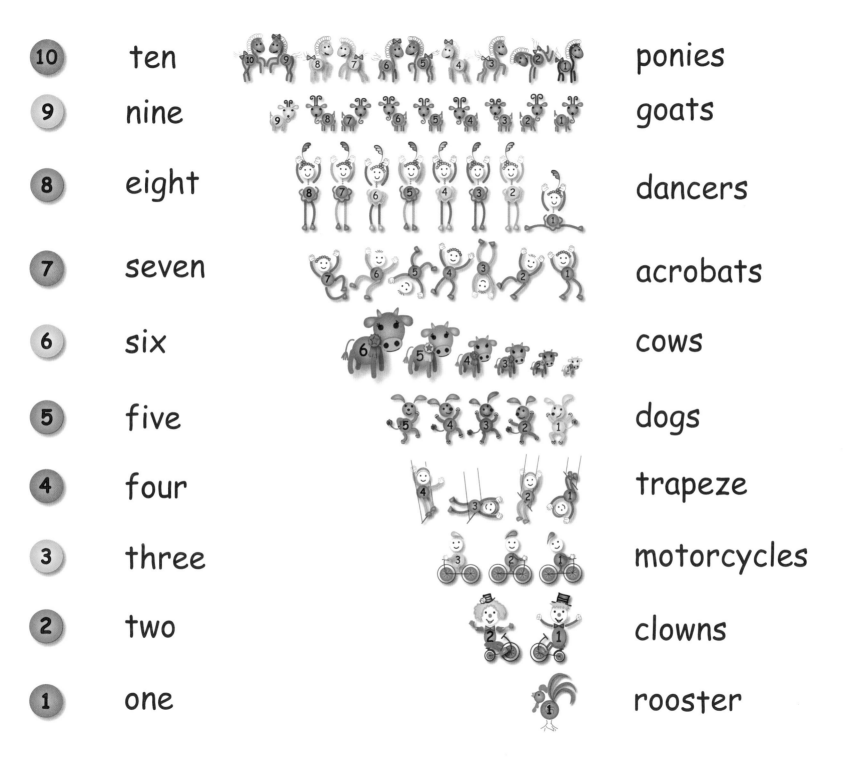

10	ten	ponies
9	nine	goats
8	eight	dancers
7	seven	acrobats
6	six	cows
5	five	dogs
4	four	trapeze
3	three	motorcycles
2	two	clowns
1	one	rooster

Let's talk like a clown!

Hello!

Bee Bee's Circus

Popcorn?   Yes, please.

Ooops!

It's okay. Would you like mine?

Thank you for the balloon.

You are welcome.

Let's share!

Yes, we can share!

There are so many things to count.

We learned to count backwards.

Yippeee!

Keep on Counting!

Thanks for coming to the circus with us!

We'll see you the next time the circus comes to town!

Bye bye!

Bee Bee's Circus

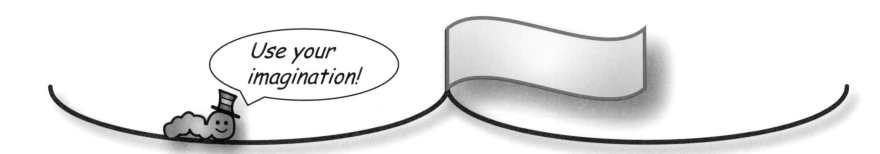

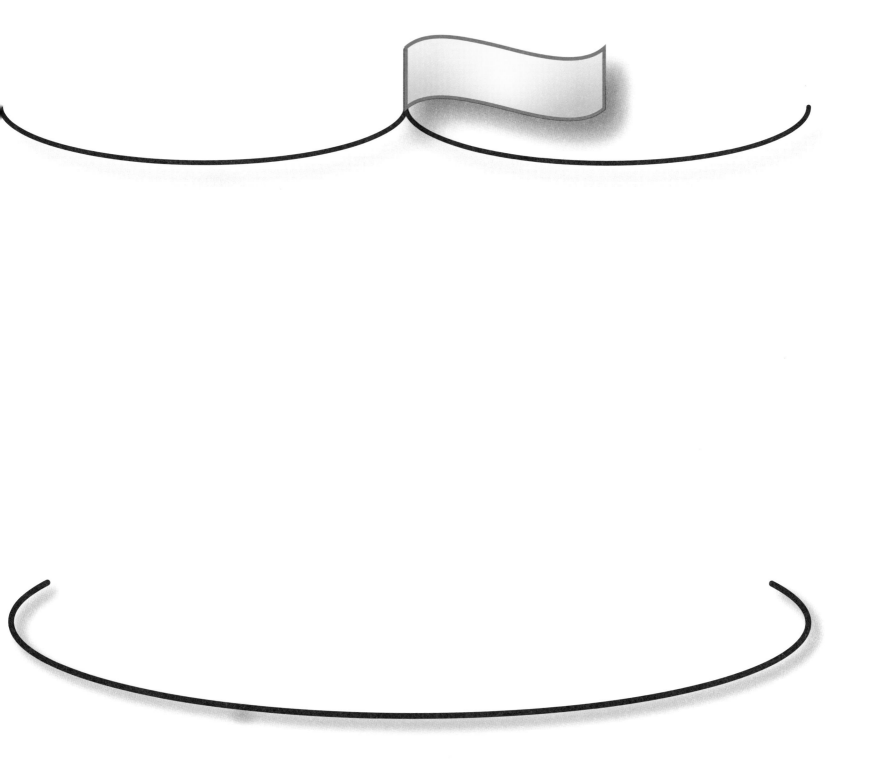

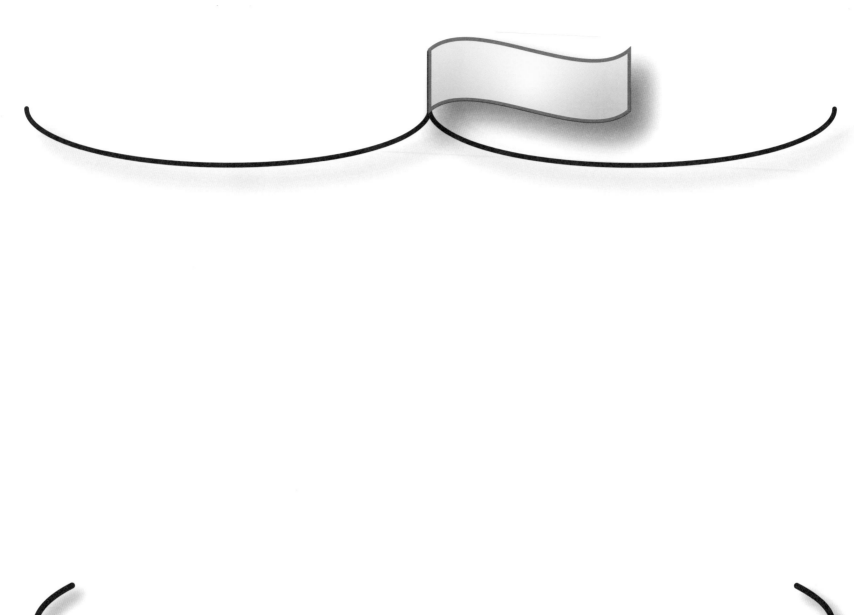

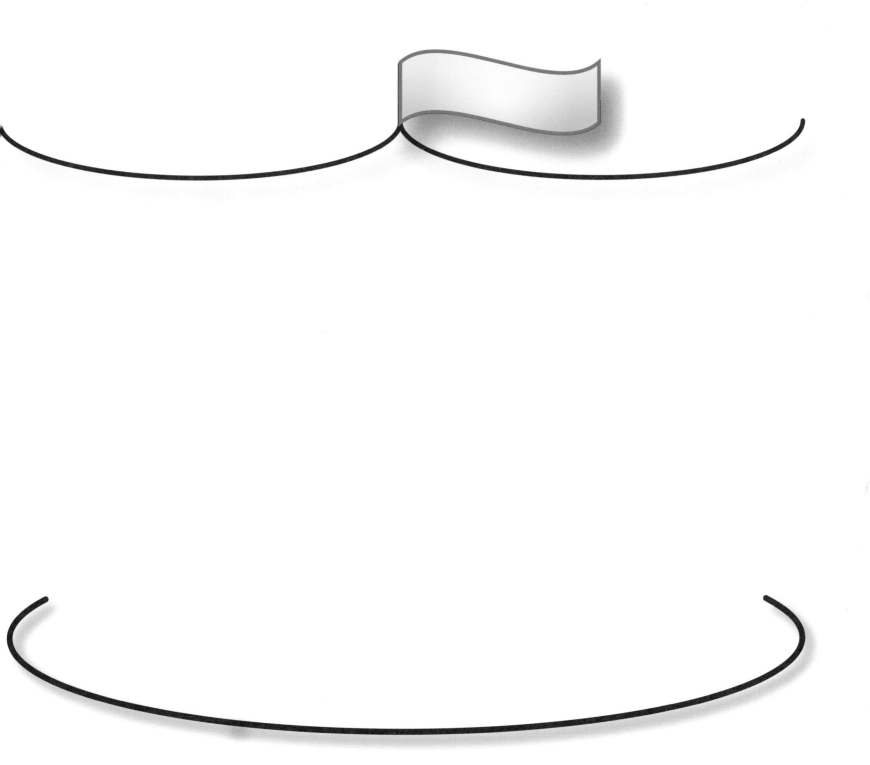

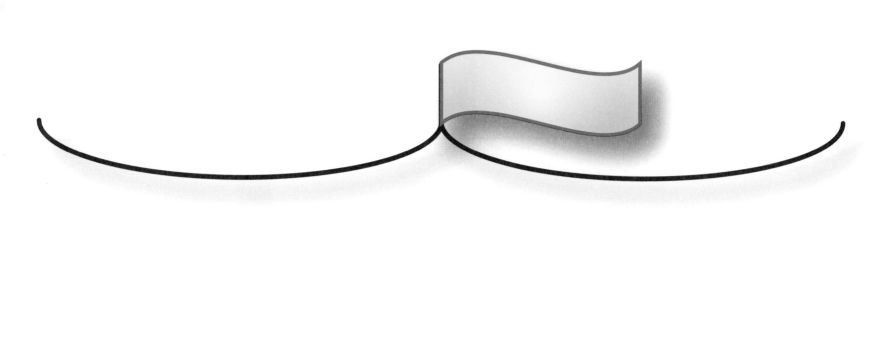

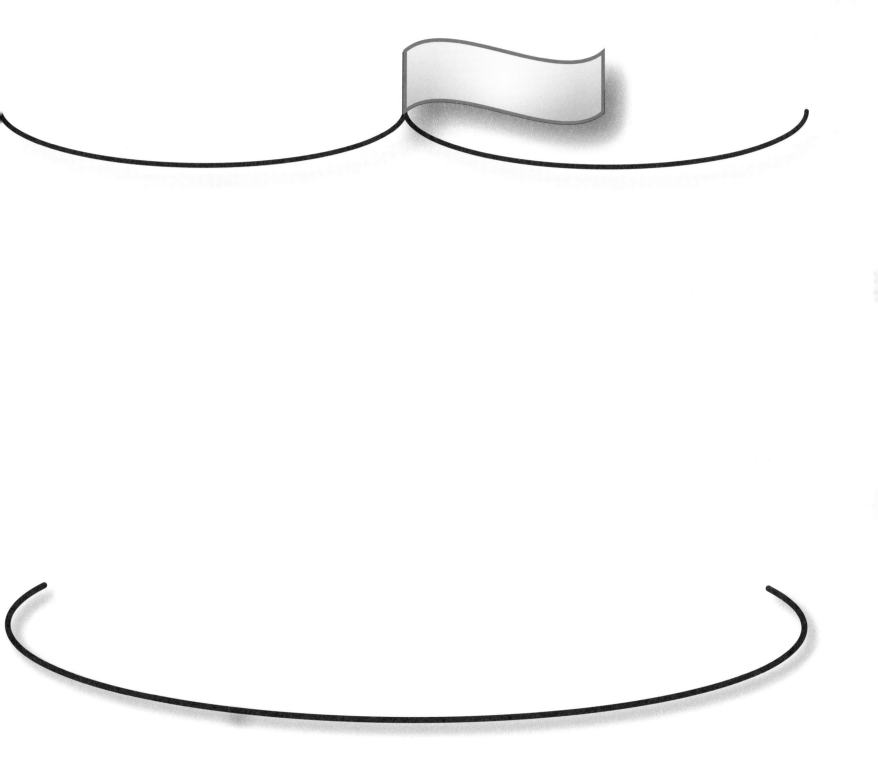

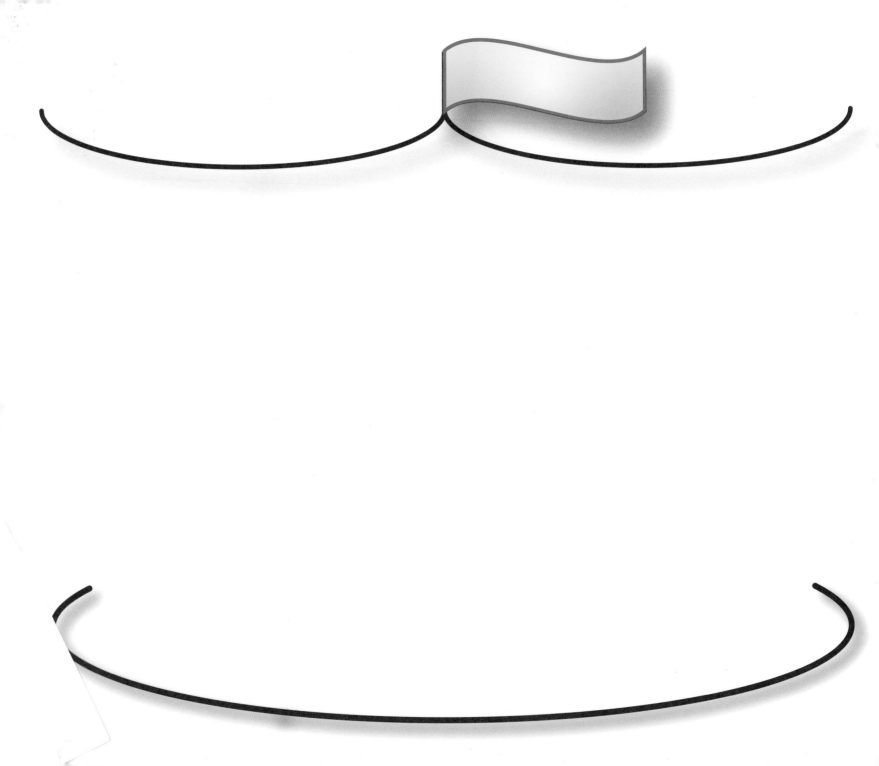